Blockchain

A Complete Beginners Guide
to the Technology Powering
Bitcoin & Cryptocurrencies

Crypto Tech Academy

The contents of this book may not be reproduced, duplicated or transmitted without direct written permission from the author.

Under no circumstances will any legal responsibility or blame be held against the publisher for any reparation, damages, or monetary loss due to the information herein, either directly or indirectly.

Legal Notice:

This book is copyright protected. This is only for personal use. You cannot amend, distribute, sell, use, quote or paraphrase any part or the content within this book without the consent of the author.

Disclaimer Notice:

Please note the information contained within this document is for educational and entertainment purposes only. Every attempt has been made to provide accurate, up to date and reliable complete information. No warranties of any kind are expressed or implied. Readers acknowledge that the author is not engaging in the rendering of legal, financial, medical or professional advice. The content of this book has been derived from various sources. Please consult a licensed professional before attempting any techniques outlined in this book.

By reading this document, the reader agrees that under no circumstances are is the author responsible for any losses, direct or indirect, which are incurred as a result of the use of information contained within this document, including, but not limited to, —errors, omissions, or inaccuracies.

Contents

Dibbly Publishing

Dibbly Publishing publishes books that inspire, motivate, and teach readers. Through lessons and knowledge.

Our Book Catalog

Visit https://dibblypublishing.com for our full catalog, new releases, and promotions.

Follow Us on Social Media

Facebook - @dibblypublishing

Twitter - @DibblyPublish

Download Your Bonus:

Bitcoin Profit Secrets

Discover the methods and techniques used by the most successful Bitcoin investors so you too can profit and succeed!

https://dibblypublishing.com/bitcoin-profit-secrets

Preface

There is a tremendous amount of confusion when it comes to the blockchain and its utility. People think that blockchains are proprietary, or a physical manifestation of a network, or something else. So the first thing we want to do in this book is neutralize all those various misconceptions and state in the most straightforward and comprehensive terms what blockchains are, how they are developed, what they do, what their risks and benefits are, and how you can make money from them.

The blockchain is not a physical asset or a physical framework by any means. You will not be able to build it using sticks and stones, nor destroy it by torching it or blowing it up! That's because the blockchain exists by means of a simple algorithm that propagates across nodes that join the network. Once the program is installed, it creates an environment and a set of processes that is called the blockchain. The nature of the blockchain is dependent upon the algorithm and the conditions set within that algorithm. In many cases that algorithm is open source, but in others it can be controlled by a developer, in which case anyone wanting to develop on that platform would need to request an SDK.

If you think that is too confusing, I don't blame you. The concept of blockchains is rather new and not initially intuitive; thus it takes some time to find the proper frame of reference for the brain to wrap itself around. So to get your mind into the proper framework, I will spend the entire length of the upcoming Introduction chapter setting up the way you need to think about the idea of a blockchain and what it can do. Then once you have that framework, it will become easy for you to understand.

For now let me just tell you why you should understand how the blockchain works, why this matters, and why the blockchain still has tremendous room to grow. You should also realize that the blockchain is an idea and a process more than the features of a particular brand.

There are also a few things about this industry and ecosystem that you should know and I will lay those out as this book unfolds. I've read the papers and the essays written by leading old-world economists and all the naysayers who say that the fundamental concept of the blockchain is false.

I humbly disagree for a number of reasons. First of all, the economics of the smart contract that rides atop the blockchain is not going to lend itself to the old-world concept of monetary and fiscal economics. Contemporary technologies, norms and practices, as well as a movement away from centralized systems, is fanning the flame of decentralized trust-neutral economies.

What are trust-neutral economies?

Trust-neutral refers to a relationship where you have zero need to trust the person with whom you are transacting with. Take for instance a simple banking transaction. You place your cash with the bank, trusting that they will hold it for you. But you have no real way of stopping them from closing up and disappearing into the night. The only way that it works is when the government stands as a guarantee of your funds. But still that is not trust-neutral. You are still trusting someone to stand as witness to your claim to these funds you place in the bank.

In a trust-neutral economy, we do not need a bank, or a government entity to guarantee our funds, because no one can use them except for their current owner. When you hold a property or an asset, the entire ecosystem is witness to that ownership—whether they want to be or not. There is no requirement of trust; it is the nature of the blockchain.

Trust neutral economics are based on self-evident transactions where no single entity is trusted to hold the truth, but rather it is distributed so that multiple unrelated nodes can check it and verify it, then keep it as part of a record that cannot be forged or duplicated. In essence, the blockchain—no matter what program is used to write its algorithm, the name of the coin that rides on top of it and who issues it—is a mutually assured ecosystem that cannot be gamed or manipulated.

Having said that, let's get into the nitty gritty of the blockchain and the features and the layers on top of it.

Introduction

As I said in the Preface, I am going to use the entire length of the Introduction to set up the preceding framework necessary to place context around the concept and utility of the blockchain.

In most cases we look at the blockchain as the core of the currency—the cryptocurrency that sits on top of it. That is especially true in cryptos such as Bitcoin, Litecoin, Ripple and so on. And that's fine because the utility of the blockchain is now dominated by the use of it as the tool to legitimize currency. But blockchains are significantly more than just that.

If you want to have a better understanding of blockchains, you should realize from the start that blockchains do not equal cryptocurrency, and in fact refer to a generic concept. In practice, smart contracts and different cryptocurrencies have varying features in their proprietary blockchains. They are nonetheless still blockchains.

Trust

All blockchains are similar because they exist to

perform one main task that has been part of human civilization ever since we started living in organized communities, and that is to fulfill the need for trust. Blockchains obviate that, and by so doing totally forego the need to create institutions of trust, like banks.

Before we begin with the setup of the existing problem, let me just add one more factor regarding the trust issue. Because of this issue for economies of scale, we created a centralized system around almost all public and private relationships. That centralized system was born of two reasons. The first was that strangers needed a common trustworthy intermediary, and the second was that the laws of equity needed to be enforced in a meaningful way. So you see how that works?

Understand also that not all actors are untrustworthy, only a few. But because there are a few and there is no prior relationship, we cannot differentiate the good from the bad and this creates risk in transactions. Risk translates to cost, and costs reduce wealth. Thus institutions were born.

Because the system is inherently untrustworthy, checks and balances along with punishment had to be instituted so that the participants in a transaction could not game or cheat the system or the counterparty involved transaction.

If you look at all the possible transactions that could

happen between potential entities there is an element of trust that is necessary, and that causes a significant amount of problems. That's the first step. The second step is the vessel of value. You cannot expect two people to come to terms on the value of two different things. Each transaction would have to be standardized.

Need for Currency

To make this clearer, let's look at a typical transaction to see why the need for currency is important in a society that specializes in various aspects. For the sake of expediency and variety, we specialize, and by specializing we are able to offer humanity better quality, variety and competitiveness. But variety brings along with it opportunity cost. If you specialize in making one thing, you forge the time to make everything else, and so you now have to rely on someone else who is specializing in another aspect.

Because of specialization, systems of bartering and exchange advanced. If I specialize in rearing goats, you can specialize in herding cows. I would then exchange my goat meat for your cow's milk. That worked great, and for the most part it allowed our ancestors to create better offerings.

But what happens when you don't want my goat's meat

but I still need your cow's milk. Now I would have to exhaustively run around the village looking for someone who satisfies two conditions. I have to find someone who wants my goat's meat and has something to trade for it to eventually get your cow's milk. That gets very complicated and inefficient very quickly. So instead of bartering one good for another, our ancestors decided that they would exchange their goods for something everyone wanted; precious metal. So this brought gold into the picture and since everyone wanted gold (because it was shiny, I suppose) gold held its own value and you could exchange it for goods and services.

There are two things you need to understand about that gold. First it has its own value. Regardless of what was printed on it, the gold itself contained value. It's like buying the fifty-dollar buffalo gold coin. It's face value follows what's printed on it—$50. But you would have to pay more than $1300 to purchase it. That's an example of face value versus intrinsic value. Gold, regardless of what was printed on it, had an intrinsic value. And so that value became the measure of the exchange. But that got fairly complicated as well, and instead of becoming a mere vessel of value, gold surpassed its use as a medium of exchange and became precious in and of itself.

And then came a second layer. You would keep the gold in a vault, and write on paper to represent the paymen., So now that paper carried the weight of the words written on it, backed by the gold in the vault of the bank. In that case the bank became the central point where all transactions occurred. The piece of paper that affected the purchase was taken to the bank and the funds given to the person bearing the document; today we call that a check.

Fast forward to the 21st century, and the gold standard is gone. Instead, the note that we use relies on the wealth of the sovereign that issues it. It is based (in the US) on the "Full Faith and Credit of the United States Government." Today the Dollar is the reserve currency of the world and it stands as the representation of the value of the US wealth. Just as the gold in the bank was represented by the paper that was written in payment back in the day.

There are a few inherent problems with this sort of fiat currency. By the way, fiats are paper currency that are backed by the government without the tangible asset that was once the anchor of the paper that was floated. Today, the currency value is determined by the political needs of the government rather than the market forces of the stakeholders.

The first problem is that the currency, to store value, must be both legitimate and unique. Which means that it must reflect some kind of value base and must not be duplicable where the same amount is spent twice. Not only is that a form of theft, it would also have inflationary pressures on an economy. This is the reason that dollar bills have serial numbers so that there is only one of each, and why it is printed securely so as to reduce instances of forgery.

But since the day when the gold standard was abolished and the currency regimes were anchored to the dollar in the wake of Bretton Woods, there have been a lot of gaps in the efficacy of the currency regime. There are inflationary forces that decay the value stored in the currency, and there exists possibility of theft. But more importantly, the nature of commerce has changed over time and we are now a society that makes purchases in absentia more often than face-to-face.

What has managed to allow us to operate online are the credit card and wire transfer industries. The problem is that these methods have become extremely expensive and cumbersome. There is no excuse in this day and age for payments to take days, when email and SMS can go around the world in a faction of a second.

The reason for this is that email and voice can be converted to bits and bytes, and that can be shuttled across the worldwide web, but paper currency cannot be torn to bits and shoved through the fiber optic cables!

Even the electronic shadow of credit card transactions needs large infrastructure to support the movement of funds thought the banking channels, making it expensive to transact. It also diminishes micropayments and microtransactions because anything that comes close to the price of a transaction is magnified in percentage. Think about that for a minute. If I bought a Ferrari from Italy and paid in cash, it would cost me $1 million and $30 in transaction fees—a miniscule percentage. But what if I purchased a $100 basket of cured meats and paid $30 for the transaction. That's 30%. Now what about buying a book for $30. Then the fee is 100%. Now what about making contributions of $5 and paying $30 in fees. You can see how it starts to get ridiculous. Of course I am exaggerating, but the point remains that the infrastructure needed to shuttle payments across the world is neither efficient nor cheap. So in a world that could move paper across oceans, we developed paper payments. But in a world where we have bit and byte transactions, we need to develop bits and bytes as currency. Enter cryptocurrencies.

Cryptocurrencies are today's iteration of the store of value. From bartering, to gold, to paper, to printed fiats, the history of currency and commerce has evolved to unite the technology and the practice of the times. And so now the world is ready for the next form of currency in the form of cryptocurrencies. We will talk briefly about cryptocurrencies here and leave the rest of the book to talk exclusively about the blockchain and the other uses that it has, as well as the potential it presents. The importance of talking about currency is highlighted because the world's first electronic cryptocurrency was based on this blockchain. Now the collective consciousness of the world has joined the two in their mind and this needs to be extricated.

Cryptocurrencies are not the blockchain. Let's be clear on that point. Just like money is not paper (just because it is printed on it). Cryptocurrencies are cryptographic vessels that contain the value that is determined by the market. The real hero in the cryptos story is not the cryptography of the coin, but rather the structure of the foundation that it is built on, travels across, and lives in—the blockchain.

Ecosystem

The blockchain can be thought of as an ecosystem rather than a network and we will look at that very soon. The cryptocurrency however, is the digital representation of the value-carrying vessel. And because it can be sent to and received via electronic means, it is a more apt instrument for the digital age.

It takes mere minutes to send payments around the world and the payments are totally secure, free (in some cases), forgery-proof, theft-proof in many cases, irreversible, and can even be used for in-person payments.

Detractors, on the other hand, will tell you that it promotes illegal trade, is used by bad actors, and that it's a bubble. But what they won't tell you is that it's a failure. Cryptocurrencies take on a different model than traditional currencies (and you should read the book on cryptos to get a full understanding of its features).For now what you do need to understand is that cryptos have all these features because of the framework they sit on. That framework is really the start of the show. It's called the blockchain.

Chapter 1

Fundamentals of a

Blockchain

So we get the idea that the blockchain is what gives rise to the possibility that a packet of data can act as a unique vessel to carry value. It is the blockchain that allows the cryptographic string that makes up the cryptocurrency, or the coin, to remain anonymous yet transparent. Further, it is able to transfer value from buyer to seller, consumer to producer, and even from machine to machine without the need for trust or centralized authority. In a world catching on to the Internet of Things, micropayments from machine to machine is going to become a necessity. This expedites exchange, reduces cost, obviates risk, and in essence transforms and transcends classical economics as we know it.

What is the blockchain?

Let's look at this using terms we are familiar with. Now,

I realize there are some limitations to the following analogy, but it is a great place to start form. Think about a bank as the central figure in financial transactions. Let's say that this was the only bank in the entire town and every single person who lives in this town had an account with this bank and kept their money there.

This bank centralized everything and kept a record of who owned what. If Alice had a thousand dollars at this bank, Bryce had five thousand, and Chrissy has two thousand, they would all have a notation in their account as to how much they had.

The bank would be in possession of eight thousand dollars. For simplicity, let us just say that was the entire population of the bank. So each account had a certain amount, but the bank had a total of eight thousand in its vault. If Alice gave Chrissy a five hundred dollar check and Chrissy promptly deposited that check at their bank, the bank would just move five hundred dollars from Alice's account and write that into Chrissy's. Now Alice would have five hundred, and Chrissy would have two thousand and five hundred. The bank would still have eight thousand dollars and they would only change the numbers in this huge ledger that they owned. They wouldn't really have each dollar marked as to whom it belonged to in the safe. It would all just be there, but the claim to ownership would be reflected in the ledger.

So in the event Chrissy gave Bryce a check for two thousand, the bank clerk would not need to go into the vault to check the balance, they would just have to open up the ledger and see what the current balance was.

As you can tell, that ledger is the key instrument in the whole bank. If that ledger is destroyed, all the cash in that vault gets frozen because no one would know for certain who owned what. They would only be able to rely on the word of the account holder and that would involve trust. But trust is not just the issue of not lying. Trust is also an issue of accuracy and perspective. With the ledger, the bank has the final say of who owns what. But with a missing or malfunctioning ledger, it gets chaotic. You get that point, right? Because that ledger is key to our discussion in block chains.

There is nothing physical about the blockchain. It is really a computer algorithm that tells a computer or a node to behave in a certain way. I am going to take you through each element of the blockchain and show you how it works and how it relates to the entire network.

The first part is the knowledge that there is no client-server relationship in this whole ecosystem. Each computer, once loaded with the blockchain client, becomes a node in a peer-to-peer architecture. Peer-to-peer architecture has been around since before the web. It's funny in that way because the original web was designed to be peer-to-peer. Each computer hosted their own webpage and you would key in the computer

IP address, then be directed to a specific port to enable you to download the information that was on the drive at the end of that port.

But in time, companies decided to centralize large computers so that no one really needed to understand how to set up a server, yet be able to outsource the task. And with that, the concept of the server farm was born. That's why we now have such a thing as shared hosting. In shared hosting, the server holds the information and when you want that information, you send a request and the server sends that information back to you.

Over time, the peer-to-peer system gave way to shared hosting, and centralized nodes of servers came into existence.

But for certain applications like file sharing, a central client-server configuration was not ideal because it would place a huge load on the server. Imagine if you had a 5 MB file on the server and 200 people connecting to the server to download it. You would burn through 1GB of transfer from the server side. That's a major cost. But on the other hand, say you had a P2P system where the files were on one computer, then it shares with a second computer, that shares with a third one, and the third shares with a fourth. Then the bandwidth on each computer on average would only be 10 MB—5 in and 5 out.

So the first thing you can see, if you have a macro view is that there is a P2P architecture that is created by the blockchain client. Once loaded and installed. It opens one port either 8332 or 8333. If you have NMAP, a network scanner, it would be really cool for you to scan a particular sector and look for open 8333 ports. Wherever you see one, you can then tell that that computer is running a bitcoin client.

Once the client is up and running it does a few things, one of which is that it downloads the full transaction ledger or part of it, and I will come back to this later. But for now, know that once you have the client up and running and you are online, you are now a node. A node is just a computer running the client within the P2P network.

You can see the full map of the Bitcoin blockchain nodes here :

https://bitnodes.earn.com/

You can see the full map of the Ethereum nodes here:

https://www.ethernodes.org/network/1

You will find the full map of the Litecoin nodes here:

https://litenodes.net/

These are just the nodes of three of the cryptos out there. In total, there are over 1200 cryptos at this point, all of which are backed by some form of a blockchain. Up to this point all we know is that the coins exist on this programmatic infrastructure that takes on the architecture of a P2P network. But there is more.

Once the nodes are set up, you have the ecosystem at the ready. This ecosystem is constantly increasing, with more people coming online to create nodes so that they can store their wallet or become miners.

So now you have your nodes and you have your client on the nodes. The next thing you have to look at is the instruction set that tells the nodes what to do. The central part of the blockchain is what is called a central ledger. A central ledger includes a list of every transaction ever created and executed within the ecosystem.

By the way, it does not just have to be currencies, it can even be smart contracts—and we will talk about the latter further into the book. But for now let's just look at cryptos, and especially Bitcoin.

The ledger is constructed line by line with each subsequent transaction and formed into a block. That block is then validated by placing another block on top of it. The more blocks there are on top of a block, the more certain you can be that the transactions in that block are legitimate.

Up to this point what you have at your disposal is just the infrastructure that makes up the blockchain. A node within the blockchain only requires three elements to be an effective node. The first is that is needs an internet connection. Then it needs storage capability, and finally it needs the ability to execute the algorithm in the blockchain's instruction.

At the base of the idea sits the notion that once a block is created, it cannot be changed. Whatever data or content that is in that block shall remain so for good. The reason you cannot change a block is because each one is tied to the block before it, and more importantly it is tied to the block after it as well. So if you want to change the content of a block, then you will have to change all the transactions in the blocks that come after that one.

If you take Bitcoin as an example, and peer into the blockchain beneath Bitcoin, you will see that there is a block created every ten minutes. This interval of time is called the "block time," which is unique to each crypto. Every transaction that occurs is captured in one block, and then other transactions are covered in the next block. In one hour, five blocks have been built on top of the original one. To be able to change the information on that first block now becomes functionally impossible because you would have to change information in the subsequent five blocks as well, and possibly the blocks before the one you are targeting!

This is part of the reason that blockchains are considered trust-neutral.

There are two things you need to keep separate in your attempt to understand blockchains. The first is the content of the data and the ecosystem that it eventually builds, as well as the features that it brings to the coin or smart contract that is built on top of it. The second is the generic nature of the blockchain and how it works. We will start with the latter first, so that you get a functional understanding of what it is and how it works.

The concept of the blockchain in essence is that data, once recorded, is stored across the entire network instead of sitting in one server. It is distributed and decentralized. When data sits in one server, you are forced to trust the administrator of that server. However when you distribute the location across thousands of nodes, you no longer need to trust any one node. The consensus algorithm of the blockchain continuously compares what is in its database and what is in the database of those it is connected to in this peer-to-peer connection. I will go into more detail about nodes, consensus algorithms and gossip protocols later in this chapter.

Difference Between Blockchain and Server

Let's compare and contrast the client-server architecture versus the P2P architecture. The client-server setup is centralized in nature. All data is centrally located and the clients contact the server to access the data or to report back with results. Every single client needs to go back to the server to get its information. Even if I was sitting in Johannesburg and was uploading a file to my document server located in Los Angeles, and my neighbor was downloading the file, we would both have to contact the server in LA. While I upload it and store it there, he would have to contact that server and download it from there. This client-server relationship concentrates the traffic around a few nodes, making the entire worldwide web inefficient.

On the other hand, the blockchain works differently than the server. The data that is in the blockchain can be data that is either dynamic or static. But the difference between static and dynamic is not the same as the difference between static and dynamic data in a server.

Here is why. In a server, when the administrator goes in to change the data, the old data is lost forever and thereby cannot be referred to in the event of need or understanding the past. We trust the server to maintain the information we give it. We trust the server to be up when we need it. We trust the administrator to keep the information as we intend. The point is that there is a lot of trust. I have found myself in a quandary more than once about whether to take an online offer, and upon purchase, the price had changed. The old bait and switch. But when I went to complain about it, there seemed to be no record of the price that was on the site which I saw.

We don't realize it, but there is a lot of trust that must be in place before e-commerce can take hold, and that is why intermediaries and institutions of trust found a niche. But the problem there is that the trust certificates and reputation management cost more, and that cost is passed back to the consumer. In the event we didn't need to trust, or didn't have to trust, then there would be less cost involved in the process. Less cost betters market penetration and the overall wealth of the system.

In overall terms, servers are centralized systems and blockchains are decentralized systems. Centralized systems are susceptible to bad actors. They are also susceptible to DDOS attacks, and even government censorship. Blockchains do not have that issue. Because they are decentralized, it would be impossible to shut down all the nodes that carry the data. Even better, it would be difficult to change the data in some nodes, in a short time the nodes that have incorrect information will be erased and the correct data will be reinstalled. It is even difficult for viruses to survive in this environment.

Blockchains are being used in many areas because of this decentralized characteristic. Even the cloud is a function of the blockchain if you think about it, but the cloud architecture is a bit of a hybrid. If you expand the ability of the blockchain, you will find there are companies like Filecoin that use the blockchain to store data over their blockchain and thereby have coins to pay those who give us drive space for this purpose. The information is encrypted and duplicated in multiple locations so that it no longer matters if your private information is in someone else's drive or across a thousand drives. That's because it has three features. The first is that it is a fragment of the data. Secondly, it is fully encrypted so that it will take an insane amount of computing power to decode. Finally, the information can't be traced back to the owner even if the administrator was willing to do so. It is a trustless system so no trust is needed that your data is safe—it is safe.

In short, there is a history of the database when it is kept in the blockchain and any changes that are made in it need to be changed on a layer, meaning you would have to issue a whole new set of data while the old set of data is present. So for instance, say you were to go to a website that advertises the price of something you want is $1. Then, when you make the purchase, the website says the item is $3. you could go back to the history of the blockchain and look at the site at the time you originally viewed it and have proof that it was indeed the price that you saw in the first place. This is of course a trivial example, but it is designed to show the power of the blockchain. The focus becomes the information itself and not the server that the information sits on.

There are numerous blockchain technologies that are emerging which look at blockchain storage. We will not go into that simply because our focus here on the blockchain is to look at the currencies that are on top of it.

There are many tokens that you could create that could sit on the blockchain, and while the token itself is not part of the security measure, it could be. But the reason that there is a coin on the blockchain is not a function of the blockchain itself, but rather a function of the app on it- in this case a cryptographic app.

Let me explain.

You see in many currencies that they limit the time it

takes for new coins to be released into the market. Depending on who the issuer is, they all have different rates of coins entering the market—they are called block times. If you notice the term they use, "block times" sounds suspiciously familiar. It should be, because it is the term given to the time it takes to create a block after solving a cryptographic puzzle. When that puzzle is solved, the miner gets the block reward in the form of newly minted coins, and when they sell that into the market, there are new coins that enter circulation.

If you adjust the block times, you can adjust the rate of creation the coin inherently has. But coming back to the coin vs the blockchain—the coin itself can be anything you want it to be. Most people misunderstand the purpose of the blockchain because they think that it gives rise to the coin. It doesn't. It only gives the coin legitimacy on its path throughout its existence. If the coin has been part of a number of transactions (the more the better) then you know that the coin is part of the network. The reason you have this is so that someone wouldn't just make up a string of numbers and claim that it's a valid coin. The coin comes directly from the program itself and only the bitcoin program determines how much reward is given. In the beginning it was 50 coins as the reward, all newly-minted coins which have never been in circulation before. After some time, that halved to 25, and then that halved to 12.5—the current rate at this point. But that has nothing to do with the blockchain. Most people confuse this and the blockchain because the concept of the coin coincided in the release of the new coins with the creations of blocks in the blockchain. It has also confused people because the blockchain was part of Bitcoin when it was first released. Everyone simply took bitcoin and blockchains to be exclusive to each other. They are not.

A blockchain is a very simple concept and here is where we are going to show you exactly what it is. So strap in, it's going to get bumpy.

The Meat and Potatoes of Blockchains

To understand a blockchain, you first need to understand a block. A block can be anything. You can put data in it. You can put an image or transaction records in it. You can basically put anything digital in there and get a block. Think of it as a box containing all your credit card transactions in a month. Then the following month you have another box that contains all your card transactions for that month, and the next month it's a new box and so on. This analogy is severely limited, but at this point I am merely showing you the blocks that build up with transactions in them. That's it. Now imagine that you tie each box to the next. This means the first box is the receipt of your first credit card purchase, and all other purchases in that month. The second box is the second month, and that just keeps ringing in until the present time.

Now that you've got that picture, let's make some modifications to it to make those boxes and transactions digital.

Now what you have is a bunch of transactions that are going on across all the nodes in the blockchain. If you are talking about Bitcoin, then it's the Bitcoin blockchain. In this case, let's just keep it simple for now and talk about the Bitcoin blockchain.

In the Bitcoin network, there are numerous transactions between wallets that send and receive bitcoins. Each send and receive is a transaction. Each transaction is recorded effectively in a public ledger and that ledger is secured and locked and cannot be altered or changed. The way it locks up is rather ingenious.

If you remember, the network that hosts bitcoin is defined by nodes on a P2P network. There are two kinds of functional nodes. One node is the kind that is passive—they just send or receive payments from the wallets in them. Another node does mining.

For the purpose of defining the blockchain, let's just look at the mining nodes, or what we call miners.

Miners are the ones that find new blocks, and when they do, are rewarded with new coins by the system.

This reward by the system is hardcoded into the the algorithm that runs Bitcoin. It cannot be changed nor altered. If you try to change it, all that happens is that you will end up forking the chain and end up having a new version of bitcoin, but you will not have changed the old bitcoin. This has happened in the past and that is why you have such coins as Bitcoin Cash and Bitcoin Gold. They forked the original bitcoin and created a new timeline.

But that's not the point. It's just information that is nice to know but not totally relevant to the present issue. So, back to our point at hand...

Miners who mine the coins are not really searching or solving new coin or finding new block. That is just the terminology and you shouldn't get confused when you hear that.

What miners are really doing is putting all these transactions that are going on in the network and packing them into blocks. Just like how you placed all your credit card receipts from one month into one box in the example above. The miners on the other hand, are competing against each other and grabbing up all the transactions that are happening and they are putting it together so that it can get verified and included into the blockchain.

Once a block is attached to the blockchain - which takes around ten minutes, then you wait another ten minutes until the next block is added on behind that one, and another ten minutes for the next block to appear after that one. In one hour there are about five blocks that have piled up on top of the first block. That means the transaction which went into the first block is pretty squarely entrenched and confirmed.

The reason you want it this way is because the design of the blockchain is meant to provide two things. The first thing is that it is designed to prevent double spending, and you do not want the owner to backtrack the payment after the payment is made. Remember the point that we talked about in the example with the server—client relationship and the way the administrator could change the data on the server, which is not possible in the blockchain. This is the mechanism that makes that possible. All transactions, once done, cannot be undone. It stands on record forever.

Let's take a look at what is in the actual block that the miners are putting together. It needs to follow a certain format, and that format is laid out by the developer of the blockchain. So the Bitcoin blockchain will have different formats and the Ethereum blockchain will have different formats. But in essence they all have similar elements that go into them. Each block has a header that contains the block version. This is followed by the hash of the previous block's header, the Merkle root, the time stamp, the target, and finally the nonce.

Just put a pin in that one for a minute. There are a few

things we need to unpack here for you to be able to visualize all that.

First of all let's try to explain what a hash is. Hash is the result of a function where you put in data, regardless of how long it is, and it returns a string of characters. Let me give you an example.

I am going to take this sentence and hash it in SHA256. You will see what that does, and then I will explain to you how that works.

"The rain in Spain falls mainly in the plain."

That's the sentence I am going to hash (without the quotation marks). And the 256SHA hash for that sentence is this:

5C2BE6976C4FB8FACC366C2DC46DEAC18292EC0
0A85EA766D6026746D83C2464

If I removed the period at the end of that sentence, the hash now becomes this:

116BDA0EA0B68B26116281A1FB0B727789CA598C
AB72D7CE543FE7EE96BA596E

Even though the alteration from the first sample to the second was merely one period at the end of the sentence, the change in the hash is completely unrecognizable.

You can try to do this as well. There is an online site that converts regular text to SHA256. You can find it here:

https://passwordsgenerator.net/sha256-hash-generator/

By the way, just as a little piece of trivia: the SHA256, also known as the Secure Hash Algorithm was developed by the NSA.

Just for the fun of it, you can hash the lyrics to your favorite song. If I took the lyrics to the "Star Spangled Banner" and hashed them, I would get this:

7FAA645319D8781CD2A26CFAEBB29DC7AB09DFC 9094E73EA99B19356D2C77ACB

It's the same length as the hash for the shorter sentences. So now you get the idea of the hashes.

Now back to the blockchain.

Once the header is in, the miner then calculates the Merkle tree of all the transactions that will be going into that block. A Merkle tree is a compression of all the hashes of the transaction ID (TXID). Each transaction that is conducted in the network, a payment for something when a coin is sent from one address to another, generates a unique identification number that is called a transaction ID. it looks like this:

8088eeadbb0c6cbc6cc87ffacc05045f50195bd3837ec
392a894465693578b57

When you take all the transaction IDs from all the transactions that you are trying to aggregate and place them in a block, you will get what is called a Merkle tree. Here is exactly how it is done:

Let's say you have 10 transactions. That will give you ten transaction IDs. If you break them up into pairs and hash them, what you are left with is four hashes. If you divide those into pairs, you will get two pairs. If you have those pairs, then you will get two hashes and that will be hashed until you get the last hash. That is the root of your Merkle tree and it represents the entire hash of all the transaction IDs in that block.

So in that block, if you now took all that information—the header, the Merkle tree, the time index and so on—you now get a bunch of data that you can process together so that you get a particular hash. So let's say this is what that looks like. I will enter the header hashes, the Merkle tree and the time stamp as follows:

B6ff0b1b1680a2862a30ca44d346d9

E8910d334beb48ca0c0000000000

0000009d10aa52ee949386ca9385695f04ede2

70dda20810decd12bc9b048aaab3147

(nonce)

The last item called the nonce is a little special. I will get to that in a minute. So now what I am going to do is put in all these numbers and get a hash out of it. But that is simple enough to do and not a big deal, right? Sure, and if I did these numbers on top (without the use of the nonce) this is what it looks like:

C808546F1FAAF31692D97D02E7B32F878E37FC8BB
7138424D29E6C2FB209E71E

But that is not good enough because if I were able to do that, then anybody would be able to get the blocks put together without any effort. So what I need to do is give them a task so that they can complete it, and give me the proof of the fact they did the task before I can accept their block.

The way blockchains do this is to either have a Proof of Work or Proof of Stake. In the case where it requires POW (Proof of Work), then the miner has to run computations in a way that the hash satisfies a requirement. When I calculated the last hash, I put in the four sets of strings:

B6ff0b1b1680a2862a30ca44d346d9

E8910d334beb48ca0c0000000000

0000009d10aa52ee949386ca9385695f04ede2

70dda20810decd12bc9b048aaab3147

And as a result, I got:

C8085
46F1FAAF31692D97D02E7B32F878E37FC8BB71384
24D29E6C2FB209E71E

But what's required by the blockchain algorithm is the addition of a nonce, which is a random number. The system wants a random number to be added to the four hashes, and that number should result in a final hash that is for a particular format. I could say for instance that I want it to start with the alphabet letter Y, or with 20 zeros. So to satisfy that, I have to randomly try a larger number of strings until I get the output hash that satisfies this requirement.

So let's say for simplicity, I use the same four lines of string as earlier but now add a fifth line where I enter a random string. I start with 1, then 11, then 111 and see what happens.

The nonce that is chosen has a certain probabilistic value and thus the difficulty of finding that number can be determined ahead of time. Like if I say I want ten zeros to precede the string of numbers in the hash, the one probability of finding that is easier than finding the nonce that results in a hash that has twenty zeros in the front end. The difficulty allows the algorithm to adjust the block times. Because the higher the difficulty to solve the hash (which is called the challenge, problem, or even the puzzle) takes longer, the algorithm is constantly monitoring the average solution time (block time). You can see that here as well:

https://bitinfocharts.com/comparison/bitcoin-confirmationtime.html

How does the system know the average time it takes to solve the hash? Well that is simple. The system monitors the hash rate in the ecosystem. You can see that here:

https://charts.bitcoin.com/chart/hash-rate

When you look at the hash rate, that tells you what the aggregate time it takes for the entire system to crunch hashes. It compares that to the average time it is taking. The system realizes that if the time falls too far below ten minutes, it means that the hash rate is chewing up the strings easily. So it increases the difficulty, which requires the miners to hash more than they did. This is the reason that the miners do not like new miners joining the pool. The more hashing power there is, the harder the solutions get. But the time is the same.

Once the miner finds the block, it means they have randomly tried and succeeded at finding the nonce that makes the hash comply with the condition set by the algorithm. And when they submit that, the system accepts the block and pays out the reward, Then everyone moves on to the next block.

As the blocks pile on top of each other in sequence, two things happen: the records they keep become a live version of what is current and this is attached to a historical record of what has happened. In a blockchain, you can trace the sequence of events that has transpired for each and every coin from the time they came into service until the minute you are checking.

This has a number of advantages, especially in use as the backbone of a currency. First of all it gives legitimacy to the coin. Each and every coin is transparently seen on the map of the blockchain. You can trace any coin and see if it has been used in the past. The more the coin has been used, the more you know that the coin is legitimate.

Whether or not a person has the right to spend the coin is not part of the blockchain; that is part of the coin's algorithm itself. There are ways to make sure that the owner of a coin doesn't spend it more than once, or does not pay two people at the same time. These safeguards are built into the nature of the blockchain, and I'm going to tell you how this is accomplished.

The blockchain does this by having verified transactions that are placed in the blocks and then mined. Once the transaction is computed, the system automatically knows who owns what, so the person that does not have ownership of the coin is unable to spend it. Because the transaction is verified, soon after it is placed in a block and then onto a chain. The moment the block has other blocks piling up on it, the legitimacy of the transaction climbs significantly. In most cases it only needs one confirmation, but if you have four or five, it makes the certainty rock-solid.

Once the block is confirmed, then the miner transmits that information to the nodes which are around it. The miner typically has between 6 to 10 nodes connected to it, and so it broadcasts the new block to those nodes. Those nodes then update their records and share the result with six to ten other nodes that they are connected to. So by that point—if each node is connected to ten nodes—then by the second instance, 100 nodes already have the data. By the fifth leap, the new information has updated a million nodes, and it only takes a few seconds for each update because the block sizes are not more than 1MB (Bitcoin). This is called the gossip protocol. Because it fairly mimics the way humans gossip. One person whispers to the next and says, "Hey, have you heard…?" So if what one ledger has is different from what the other has, then they compare, and the one that is deficit in the new information copies it on to his node.

This is also how all wallets work. They do not just look at what is coming to them and then remember that information. Because they hold all the data to every transaction (in a full node), plus all the data and information of the coin throughout its lifetime (in the shortened node), the wallet gets its data from the node and looks at the history. It then realizes that one or more (or a partial amount) of coins belong to it.

But then that should raise the question of legitimacy of the information. There are two risks to this and there is one way to fix them both.

The legitimacy of the transaction is also protected by one more layer of security in the blockchain.

The mechanism of how the coin is owned and the public and private key pair are not part of the blockchain. They are part of the mechanism that defines the coin. So imagine a planet like the earth where you have the core, the mantle and the crust. The blockchain would be the core. The currency or the token, like Bitcoin, would be the mantle. And the apps or DApps that use the blockchain or the coin would represent the crust.

The whole idea of the blockchain is to remove the need to arbitrarily trust a central institution (which is better than arbitrarily trusting the counterparty). You know, make sure this is no longer part of the transaction equation. This reduces the cost of transaction and increases the the speed of each transaction and their incidences.

So that, in essence, is the nature and function of a blockchain.

As a recap, you have the DApp (Distributed Application) which resides on the top layer. Beneath that you have the coin, and under that you have the blockchain. When you separate the functional aspects of the cryptocurrency into its layers, you start to get a better picture of how it all works. An example of a DApp is the wallet that you use to store the coins. The wallet listens to the blockchain to update itself and understand what data is coming to light in the blockchain, especially data that is relevant to its wallet address.

The final thing you need to understand is the feature of the blockchain that prevents fraud and forgery.

Consensus Methods

This feature is called the consensus algorithm and it is contained within the blockchain's root algorithm. It is one of the core functions and is related to the gossip protocol in a way that it not only tells the next node what it knows, but also checks to see if that node has anything to add.

The consensus algorithm is always taking information that is out there in the node population and attaching it to the previous block already stored in its own memory. In the event someone tries to make their own block (or alter a block to include an unsanctioned transaction) and pass it of as a legitimate block, the few nodes around the offending node may accept the block for a few moments. But then the moment they receive word of the other block which has been legitimately accepted by the general network population—and that information gets to them—they will now realize that there are two blocks in the system vying for the same space in the chain.

Since there cannot be two blocks in the same location, the nodes will then wait for the next block to be processed and 'found.' The next block will only come from one of the two because it is very likely processed by someone other than the offending miner (node).

For the block to be affirmed, it has to gain the consensus of 51% of the nodes in the network. The only way the node in the network will affirm the block is if it is part of the longest block. That is the only way the nodes (or rather the algorithm in the nodes) will reach a consensus. If there is a discrepancy as we mentioned earlier, then the nodes will wait until the next block is attached, then the next, and then the next one—until it starts to see the pattern of blocks that are slowly increasing. It will then erase from its memory the offending block. The entire network will reflect the appropriate transactions in the block at that point.

This is the reason you are advised to wait to ship or provide the service as agreed after receiving the payment. It is not until there is confirmation and that other blocks are built on top of the block that contains your transaction, that the transaction is considered complete.

Chapter 2

Public Blockchain

There are three broad categories of blockchains. The Public Blockchain is the first one we will look at in this discussion because it is the one that is widely known at this point. It is also the one that came into being ahead of any other blockchain in the industry. The public blockchain is the underlying mechanism that accompanies Bitcoin and it is the one that served as the original template for not just the algorithm, but also for the concept.

As you advance your knowledge and understanding of the blockchain you will start to see that indeed the blockchain is separate from the cryptocurrency—or at least it can be. A lot of confusion has reigned in the space of cryptos and block chains over time because they seemed to have emerged in the collective consciousness of the world at the same time. But as you have now realized, they are actually different. Because they are different, they can be used to affect other kinds of transactions as well, not just ones that involve payments.

There are also things called smart contracts and smart transactions. It is possible, too, to use the blockchain as part of a storage system. Just as you would have storage in a server that is centralized, you also have storage on a blockchain that is decentralized.

A public blockchain is exactly that—it's public in the sense that anyone can participate in it. Anyone can monitor and confirm transactions, as in a regular node in the network. Anyone can be a miner to create blocks and do the computations necessary for the proof of work. In fact anyone can keep an eye on the open-source codes that define the blockchain. It is, or at least it is trying to be, as decentralized as possible in absolute terms.

Any attempt to block usage for certain users who want to be a part of it, can be considered a form of centralization. The definition of centralization here has to be pretty wide, but it still does fall within the ambit of less than total and absolute transparency and availability—just like the Bitcoin blockchain.

There are no barriers to entry in a fully public blockchain, and it can be easy to adopt. For instance Bitcoin runs on a fully public blockchain. This means you can still set up and run your own node and wallet, see all the transactions, and confirm transactions without even owning a unit of the coin. In fact, you do not even need to run a node; you can go to the relevant websites and see the block time, hash power, node map, transaction details and everything that you wish to know. You can then build applications on top of it. If you want to mine it, you are also free to do so. However, market forces have made mining an expensive proposition.

But that is not something that can qualify the blockchain as being totally, or even partially private.

So the elements of a public blockchain are that it is decentralized and liberal. It allows anyone to participate regardless of location or intention. It also allows anyone to purchase the coin, mine it and build apps on top of it. There are no prerequisites to any of these, except of course if you plan to mine it, then you need to have the relevant equipment to do it effectively.

The concept of blockchains was born out of public utility, and its initial introduction to the world via the Bitcoin network was designed to be all-inclusive. From the open source coding from the blockchain, to that for the original wallet, nothing has been shrouded in secrecy. The only thing that is kept private is the actual identity of the participants, and even that is because the data is not needed to participate. No one holds this data in any form.

The only way someone can identify your transaction is if you give them a payment from one of the bitcoin addresses that are related to other accounts, and then they can see the payments that you have made from that address. To overcome this and maintain secrecy, savvy bitcoin owners typically route their payments across multiple addresses and then split this up to avoid detection, before finally making the payment that they want concealed.

Remember that setting up wallets is free and so is getting a bitcoin address. There are a total of 2^160 possible addresses — that works out to being 1.46 x 10^48, or 1,461,501,600,000,0000,000,000,000,000,000,000,000,00 0,000,000 possible addresses. That still gives each person on the planet possible access to use 1.46 x 10^38 addresses each. So not to worry, there is no possibility of running out of these anytime soon.

The larger the blockchain, the more robust it becomes. Why? Because the more fragmentation that occurs in ledger existence, mining activity and confirming activity, the less chance one or more bad actors can have to alter or change history or the data within the blockchain.

In this respect, public blockchains have tremendous benefit over the private version of the blockchain. But then again, the private version serves a unique purpose and we will look at that in the next chapter.

Now let me be clear about one thing. Blockchains are not proprietary algorithms. They are concepts. You do not need to use existing algorithms or source codes to build your own blockchain. There are two very simple avenues that you can take. The first is, you can fork any open-source blockchain and make the edits you need.

The second is that you can write your blockchain in any language you choose.

Just Google it or go to GitHub and you will find a large number of blockchain codes in Python or C++, or any other language that you are comfortable with.

Bitcoin and Ethereum are two of the most popular coins that are built on top of a public blockchain. There are others, but you get the point on it and you understand the public blockchain enough to recognize it at this point.

Chapter 3

Private Blockchain

The Private Blockchain differs only slightly in construct and architecture to the public blockchain. The difference is not central to the way it operates, but on who has access to it and in what capacity.

Think of it this way. Let's say the corporate governance laws change (this is really to illustrate the point—not the intention to suggest the need to change corporate regulations). Each corporation is required to keep their ledger of revenue expenses and assets on a blockchain. What would happen is that the company would only allow a few nodes to participate in the ability to add to the ledger; remember, they cannot change the ledger once it is in the block.

On the other hand, anyone who needs to have an accurate snapshot of the funds can do so by extracting the historical data whenever they desire to. The same goes for viewing the movement of funds across borders. You could write a short app to extract data from the Bitcoin blockchain and you would be able to see the inflow and outflow of all funds from within the US to points outside the US, and vice versa. All it takes is a little app to connect to the blockchain, and getting this is a breeze. But the point is that the same could be done for the company. Now if the company is a publicly listed one, then these accounts are supposed to be within the public realm anyway.

So in this case, the bookkeepers and accountants could place the accounts on the blockchain, and the managers/stakeholders, or auditors would be the 'miners' who pull the transactions and make them into blocks that are considered approved. Once that is done, the records of revenues, expenses, assets and liabilities are all set in stone, until the next event gets updated in the ledger and the machinery runs once again.

This would be one way to use a private blockchain. No one has the freedom to go in and change the data, but everyone has the freedom to view it. In some cases the administrators could also secure the data in such a way that only certain people could see it, and then that data would not be designed to go into the public realm until and unless a certain event occurs. For instance, you could have a corporate blockchain that has the bookkeeping we just talked about earlier. Then when the managers approve the data and they are put into place, they finally come to be released in the event that the numbers are audited, when the data must be released for public view.

This brings us to the element of the smart contract.

Smart Contracts

Smart contracts are not the exclusive domain of the private blockchain. You can institute smart contracts in any you wish, but you should look to do them on blockchains that are Turing complete. The Bitcoin blockchain is NOT Turing complete, but the Ethereum Frontier is. That is the reason smart contracts prefer to be placed on the Ethereum Frontier blockchain.

What is Turing complete? Well it just means that the platform is able to compute and perform tasks that any Turing computer is able to perform—tasks pursuant to the algorithm it is fed. And what does that mean? Simple. It means that the scripts that are uploaded to the blockchain are able to execute conditions, loops, and so on just like any program, so that when you have these scripts or apps they can automatically evaluate an event and then perform a function.

So let's say for instance I run a public company and my shares are in the form of tokens as part of the blockchain in my company. If I promise my shareholder that if the company's profits are above $1 per share at the close of the financial year, we will declare a bonus to each coin holder.

This is written into the smart contract and when it comes time, the script will monitor the ledger. As soon as the earnings are released, it will look for the earnings number. If that's indeed above one dollar, it automatically issues a coin to each coin holder as per the contract. There is no further tampering or tinkering by anyone. It is automatically done when an event is reached.

These are smart contracts and the potential to run these is extremely lucrative and wide-ranging in the real world. This is the reason blockchains that have full scripting ability, like Ethereum, are so popular. Bitcoin's blockchain does have limited scripting, but they are so hamstrung that you can't really do much with it because that is the way Satoshi designed it for a certain measure of security.

Just to be clear, Bitcoin has some scripting ability, but it is not Turing complete, so its ability to write robust contracts is not present.

Although smart contracts can be written in private as well as public blockchains, it finds tremendous use in the private scenario especially because of the kinds of relationship these closed groups may have.

A private blockchain is considered partially centralized because part of the ability to alter the ledger is limited to a few persons only. That centralizes the blockchain by default and turns all the other participants into data gatherers and decision makers. Or it makes them a stakeholder but not the financial manager or the operational managers.

With blockchains, there can be added transparency while maintaining secrecy and privacy. For instance, in public blockchains I could use my bitcoin to purchase an item that is totally transparent and anyone could see the purchase, but they would have no idea who the person behind it is. That anonymity has its benefits.

For a private blockchain, the anonymity is not some much the factor as proprietary. You may not want your competitor to see how much you are making before the numbers are made public, but at the same time this kind of transparency gives shareholders and stakeholders significant insight into what is going on with the investment.

If you want to set up a private blockchain, there are a number of ways you could do it. If you understand C++ or any form of C programming language, then you could take a copy of the open-sourced Bitcoin blockchain, make the necessary changes to the code and deploy it as part of your private blockchain.

The other way you could do it is to use the Oracle framework and design your own blockchain based on the Ethereum framework. Whichever you chose, you will find that there is a significant number of programing required, even if it is only to alter the parameters that define the blockchain. In fact you can even create your own token or cryptocurrency. As long as you've got the blockchain and the parameter that conform to your vision, then the rest is pretty easy.

You can find more ideas in Chapter 5 on Blockchain Development.

The fundamental aspect of blockchains that you want to remember is that it is more decentralized, authenticated, trust-neutral, and high value-to-cost ratio than other mass-deployable technology presently in existence.

Chapter 4

Federated Blockchain

The last category of the blockchain that we see developing over time is still pretty much based on the full blockchain, except there is a slight difference in the way consensus is built. As a recap let me just remind you that the benefit of the blockchain is that it is decentralized, and trust-neutral. That means that it obviates the need for intermediate institutions to provide the element of trust. Think about that for a minute. The first catastrophic element of this is a major part of a bank's business. The bank is typically the center of the financial transaction and it allows for parties to understate transactions because there is a bank in the middle that allows them to take comfort that a trustworthy party is involved.

But in federated blockchains that trust is not necessary, and this results in three direct consequences. More relationships can be entered into, increasing the frequency of business and thereby increasing the potential and probability of profit. The second is that it reduces the cost of the transaction while still maintaining a high level of trust in the relationship. Third, it opens up new and vibrant ways to conduct new business and micro business, where the transactions are numerous and the cost or the profits are small—but add up.

The blockchain also provides security in two ways. The first way provides the security of maintaining the data integrity. Secondly, it provides security of the data itself against attack. In a blockchain, data is immutable. Bad actors can't hack the system and delete the data. Let's see how that works and why it makes currencies built on top of blockchains secure and inexpensive.

The feature inherent in the blockchain is the consensus algorithm that we talked about in the earlier section. This is the main piece of the puzzle that allows the network to maintain its data integrity. Look at it this way. If a bad actor tries to change the data that is already in the blockchain, there are two things that he would have to do. First, he would have to change the data itself, which will inadvertently alter the block. The moment the block is altered it no longer will fit in with the block that comes after it. Remember the nonce and the hash. So to change something that has already been confirmed into a block requires tremendous resources to change - even if it is just one block. Because that requires even more resources to change the block that comes ahead of it and the subsequent block after it. This alone is no longer viable unless new computing ability enters the market where creating 25 terahashes is a trivial matter. Such a level of computing power may exist in quantum computing, but that is not commercially available just yet. When it does get to the market, it will still require a tremendous amount of real time input to change the course of the current transactions. All in all it is not a viable proposition.

The second form of security it provides is in the event of a virus. Nodes that do not conform to the network standards are not allowed to broadcast, so when that happens, the node is disabled and it can only accept data instead of write to another node. At the same time the port 8333 where the information comes in, uses a sector of the hard drive that is encrypted. In Bitcoin the data is transparent, but the thing that holds them together is the SHA256 encryption that changes drastically the moment even one bit of data in the original file is changed. So the likelihood of viruses and trojans passing from one node to the other is slim.

But the possibility still exists, however small, that someone could figure out a way to transmit malware via the network. When they do, the safeguards are in place, none of which include antivirus packages will isolate and the node. New data can then be written over it, rendering the virus inoperable.

This makes for a robust system. However, there are areas where it can still be improved.

Areas Where Blockchains Still Fall Short

Not everything in the blockchain world is a bed of roses. There are still significant areas that need to be improved upon and problems that need to be solved before the widespread adoption can be contemplated. At this point in time we find pockets of users, the largest pocket undoubtedly being Bitcoin. These smaller pockets allow the limitations of the blockchain to go seemingly unnoticed, but there will be a time in the near future when we start to see this effect have a real impact on the layers above the blockchain. Here are just some of the things that you need to watch out for and find solutions to.

The first, as it applies to cryptocurrencies layered on top of the blockchain, is the scalability of the system. For comparison you need to understand that payment systems like Visa that process payments around the world, can handle at least 2000 transactions per second. By contrast, PayPal handles about 8% of that, averaging about 155 transactions per second. That may not seem bad, but that is in a world still pretty new to e-commerce. Imagine what it would be like a decade from now when online retail goes through it explosive growth. In 2016, 1.86 Trillion in US dollars was spent around the world in e-commerce. Every single one of those transactions had to be executed via some form of payment processing gateway like Visa or PayPal, or some form of cryptocurrency. Just as a side note, you can readily see that the payment gateways make a tremendous amount on fees. So understandably, they're hesitant to endorse any technology like the blockchain technology and cryptocurrencies, seeing them as disruptive technologies that will eventually displace them. Hence, there is a lot of propaganda and government lobbying to rule out cryptocurrencies.

Ok, back to the payment processing industry. In 2021, that spending number is expected to reach Five Trillion US dollars. For that kind of quantum, robust systems are needed to handle payment processing. The current speed of transactions Bitcoin is able to handle is 5 transactions per second. Remembering that Visa is 2000 transactions per second, do you see what a bottleneck this could be? The next iteration of blockchain need to devise a systemically faster way to process payments.

We can almost see that Bitcoin and the blockchain that it sits on, have significant issues when it comes to transaction times. It needs to get to a point where you are able to transact at least a million transactions every second for it to become viable and relevant in the 21st century, especially post-2021. At the moment, Ethereum's blockchain is able to handle 100 transactions per second—a rate that is twenty times greater than its predecessor. As it stands, Bitcoin is already starting to feel dated.

But the thing to also note is that the folks at Bitcoin are looking into ways to handle the transaction rate, and this is working into some viable BIPs (Bitcoin Implementation Proposals). Only time will tell if they are successful, but if they don't come up with any, there are other blockchains that are coming online which do indeed have the capacity to give Visa a run for their money.

A further problem with the Bitcoin network is that there is now a fee for each transaction. I am not upset that there is a fee, but what is cumbersome is that the fee is not implemented in a way that promotes the use of cryptos. In fact, it makes micropayment unviable. Here is what he fee on the Bitcoin blockchain looks like.

They charge based on the size of the transaction in bytes. It's not the amount of BTC that you remit or receive. It is the size of the file that you send. So if you have a long note in the transaction, then the size of the file to be included in the block changes the price—at this point in time, 20 satoshi per byte of data.

Ok, let's unpack that before we go on. What is a satoshi and how to do you look at the size of the data in your transaction?

Here is how it works.

1 Satoshi is one hundred millionth of a coin. In other words, 1 satoshi is 0.00000001 BTC. So 20 satoshi per byte works out to being 0.0000002 BTC per byte. A typical message is 226 bytes. That means the fee, at minimum, is 4520 satoshi or 0.0000452 BTC, which is $0.45 at the current exchange rate.

There are two inherent problems with this fee schedule, even though it is still cheaper than any other service I can think of. The first is, this is the minimum fee required. Miners don't look at the minimum fee, they look for the ones that are willing to pay the most, and those are the transactions that they will process first. Miners get their income from two sources. One of those sources is from the new bitcoins they get as reward when they process or 'find' a block. The second is that they get all the fees that are accumulated from all the transactions in the block. While it is important for the miners to make money, the fee is also subject to one other fact: the miner is able to choose which transaction he wants to include in his block.

If the miner only chooses blocks with huge transaction fees, then what happens is that the ones that have those high fees tend to take a longer time to become part of the blocks. Ok, so this is not such a big deal, but the one thing problematic is that as the price of bitcoin goes up (almost $10,000 at time of writing), it means that the fee is $.50.

When BTC was $5, that fee would be less than a penny. And so macropayments and transactions were viable. That was one of the great pulls of the crypto. You could send small payments anywhere around the world. Now if you want to send $1, it will cost you 50% in fees to only do that. What happens when you want to send a penny?

This need to be reconsidered in the next iteration of the blockchain and how to democratize the block inclusion criteria. At the moment, the transaction is not treated with neutrality as it should be. I mean, after all, we are creating a platform that is amazing in every other way; a little more thought into how to charge for it should be happening!

The third area of the blockchain within the Bitcoin universe that needs improvement is the Proof of Work formula they are using. Ethereum has made concrete plans to move away from this and move to a Proof of Stake formula. A proof of work formula is a novel idea and suited the Bitcoin ecosystem when it first came online, but today the computational proof of work it requires is one that has a detrimental carbon footprint.

How? Let me explain.

I am not going to get into the full block creation

process, since I have talked about it a little further up. But what we need to keep in mind to make the current topic relevant, is the fact that massive quantities of electricity are used, generating huge amounts of heat, to compute the hashes that satisfy the difficulty rating.

In short, the entire bitcoin network uses 30 terawatt hours of energy to keep the network going. This is the computational proof of work that it must compute, using the processing power of the computer, or more typically the GPU or the ASICs that are more efficient in hash rates. Whatever the device or the equipment, the power consumption is a lot more than people realize. And the issue does not end there.

The way it is set up if you recall is that it needs to keep the gaps at 10 minutes. So with that in mind, let me run a simple thought experiment by you. Let's say you have 1000 machines taking up 1 gigawatt of power and the algorithm decides that with so much hashing power it needs to raise the difficulty level. This means it takes more computations to get to the desired puzzle solution. That results in the ten-minute block time. That is fine. Now let's say people find that because the price of bitcoin has gone up, more people find it lucrative to mine the coin and so now the hashing power doubles. Now you have 2000 machines taking up 2 gigawatts of energy. But the moment the hashing power increases, the system detects that and increases the required difficulty to keep the block times at 10 minutes. So now what is happening is that the entry of more miners increases the aggregate power consumption of the entire network, but the block times stay the same by design.

So what effect does this have? Well, two effects come to mind. First, the more people learn about mining and joining mining pools or set up their own mining rigs—that consumes more electricity overall. If the hashing power increases by tenfold, the difficulty will also be increased and the block times will revert to their 10-minute intervals. All this increases the usage of current. It would be ten times as much as at the beginning of this thought experiment I've shared with you.

The problem with this is that many countries subsidize their power. Many countries also do not use clean fuel for power but instead use coal-fired plants or other non-greenhouse fuels, emitting gases. On average in total to this point, the bitcoin industry alone sucks up 30 terawatt hours of energy per year and leaves a carbon footprint of 740,000 metric tons of carbon dioxide.

That is a catastrophic amount. Something better in terms of proof of work needs to be instituted because at the moment the market price of Bitcoin is driving a large number of miners to come in for a piece of the action. The only thing that is doing is increasing the carbon emissions and raising the total energy consumption. The current price for carbon credits is approximately $15 per metric ton. For 740,000 metric tons across the Bitcoin ecosystem in a year, that works out to be $11.1 million. Maybe the miners could find a way to purchase these credits if and where applicable. That's just for Bitcoin though. There are more than 1200 other cryptos in the world today, and none of them come close to the kind of volume that Bitcoin miners generate—more than 25% of the total market value by exchange transaction, and significantly more than that in terms of mining activity. Just to give you an idea the total hashing power, the Bitcoin blockchain ecosystem has 25 million terahashes per second. The closest second is Ethereum, which has 250,000 Gigahashes per second.

Energy consumption and carbon emission is not something that most miners and cryptocurrency enthusiasts talk about. And admittedly it is not something that one thinks about when mining is so lucrative compared to the cost of power. In the future when power is in abundance because we have managed to tap solar, geothermal, nuclear and wind power (forms of clean energy), then it will make more sense. But for now, better proof of works are needed to drive the blockchain effort forward.

These are the main problems of the bitcoin blockchain that can be easily rectified in the future, and we will talk about that in the next chapter on Blockchain

Development.

Chapter 5

Blockchain Development

In the first half of this book, you saw the tremendous benefits that blockchains have in certain areas of human interaction. In the last chapter you saw how even with all those benefits there are still certain areas of the concept and execution that are significantly deficient in overall performance. Also, significantly short of its potential.

What that means can be unpacked into three baskets. First, there is tremendous potential in decentralization, a concept that has been left outside the realm of discourse simply because no proper mechanism has existed to replace centralized institutions as the arbiter of trust. This potential is merely the tip of the iceberg because the uses are wide-ranging.

Second, the difficulty in moving forward is also tied to a severe level of misunderstanding of what the blockchain is or isn't. There are large swaths of people, civilians and commentators alike, who believe and expound that blockchains are cryptocurrencies. They are not. Cryptos are merely one possible use of blockchains. The mistake in placing them both under the same heading creates tremendous problems in adoption. Why?

Because somewhere along the line, bitcoin has become attached to the nefarious actions of bad actors. Thank heavens that this reputation is slowly being diluted—witness the fact of the rising market price. But major economists still see cryptos as a hazard rather than a plausible alternative. It's not. Blockchains are no different from knives. You can use them to perform life-saving surgery (in the form of a scalpel) or you can use them to gut a deer with. The knife is a knife, is a knife - how you use it determines its shades of good or bad. Same with the blockchain. It is just an architecture that offers decentralized operations and a way to do away with inefficient, centralized institutions. But the bottom line here is that you have to first divorce bitcoin (as in the actual coin) from the blockchain that powers it beneath the hood.

Finally, the concept of trust-neutral relationships can fuel significantly more powerful relationships and opportunities rather than relying on trust intermediaries. Payment processing is only one area that can be democratized, there are numerous other areas as well, and all these need to be handled in three steps. Step One: the need to be visualized outside the box. Step Two, the need for redesign in terms of hardware and, Step Three: regulations need to be torn down to reduce the kinds of things that currently inhibit the benefits of the blockchain.

There is one last area that needs to be looked at, and that is law. Lawmakers and enforcers need to take a fresh look at blockchain without the taint of looking at a currency that is built on top of it. It indeed takes time to get to understand a new paradigm, just as it took some time for lawmakers to understand the internet when it first gained prominence.

Mistaking cryptocurrencies and all the insinuation of nefarious activities of a secret medium of exchange, for the utility of the blockchain beneath it is a mistake. It would be like mistaking pornographic websites in the early days of the Internet and saying that the Internet equals porn. Neither does the Internet equal porn nor does cryptocurrency equal the blockchain!

The blockchain, like the Internet, is merely the platform. The latter democratizes information, the former decentralizes it.

There are three areas of the blockchain which will see rapid advancement in the next five to ten years, and this is the space that you should be looking at if you intend to take on the blockchain as a platform to launch a new business. The days of Web 2.0 are coming to a close. The next web platform is in search of a technology and architecture that will move its ability past the inherent bottlenecks of the old system.

The first major area that is obvious, and already being advanced, is the smart contract. Ethereum has spearheaded this technology for that purpose, and just as Bitcoin gained notoriety and brought cryptocurrency to the forefront, Ethereum is doing the same for smart contracts.

What is a smart contract?

This is the kind of contract that is self-executing without the need for third-party involvement, and the benefits of the contract automatically accrue to the parties in the contract. Let's look at a simple example.

Imagine a car rental agency where you book your car online, provide them with payment information using the crypto-based Ethereum platform. But the amount is not released; it is placed into a binding contract. The contract is for you to do one of two things: pick up the car, or cancel it before a certain time. The contract is placed within a condition in a short program. No payment is made but you are given a four-digit code, the plate number of the car and the lot where it is parked (the exact lot is known because the sensors on the car know exactly which lot the car is parked in). When you arrive at the airport, you go straight to that lot, find the car, and on the door is a keypad where you key in that number. The car unlocks, you have taken possession of the vehicle, and the crypto agreed to in the contract is transferred to the rental company. You get to drive off with the car. No standing in line. No last-minute changes, no hassles. No problems with a credit card, and the contract that was pre-agreed upon gets executed effortlessly and without any form of human intervention.

Think about that for a minute. Think about the savings in terms of labor and infrastructure. This car rental company can be located at any parking garage without having the need to pay for an office or people to staff that office. Contemplate a company like Hertz and the amount of money they pay in rental at airports across, let us say, just 200 major airports. Now remove that entire cost from their expenses.

That cost savings will result in cheaper services across the board. Not just in car rental, but also with individual businesses on a much smaller scale.

Think of the blockchain as part of the architecture for distributed storage services. There are already efforts to develop this beyond just the file storage that is available today. But the development of the decentralized blockchain to be used for storage will result in a robust distribution of parts of files that someone has.

Take for instance the fact that there are hundreds of millions of exabytes on personal computers and laptops around the world. Imagine being able to take advantage of all that unused space to distribute a file stored in a decentralized way. More decentralized than cloud servers. You see, the problem with cloud servers is that they are moving in the wrong direction according to the future. While the rest of the world is moving towards decentralization, cloud storage is partially moving to centralization. Even though the cloud servers are segmented and fragmented, they are still concentrated in nodes that are in silos. If you use storage on a blockchain and utilize BitTorrent technology—a combination of blockchain, encryption and decentralized features along with BitTorrent downloading ability—you will have bits and pieces of your confidential file encrypted. Plus, they'll be broken up and stored in multiple locations so that when you go to download them off the blockchain, it pulls the data from multiple locations and decrypts them for your use.

The benefit of this is not just the ability to store data offline. Having blockchain storage will also allow you to keep your documents secure so they won't be able to be erased or destroyed. Imagine if your hard drive gets fried, or if your server is subjected to a nuclear pulse (in this day, anything is possible) your data gets erased. But what happens when you have copies of your data spread across multiple ledgers across the entire global network? The chance that your data gets destroyed or changed becomes extremely low.

The other development that needs to happen with the blockchain is that some sort of proof of work or proof of stake be devised and developed. The issue is really about spending something in return for something. In the case of the POW that Bitcoin uses in its processes for block creation, it requires huge amounts of computing power which takes a tremendous amount of energy to accomplish. The future development of blockchains needs to make sure that the energy-dependent nature is something less intensive. Ethereum, for one, has decided to use POS in its computational work. POS is proof of stake where the owners of Ethereum coins will be allowed to mine the blocks, and then the probability that are allowed to mine would be the ratio of the coin ownership to the ownership of the pool in general.

The blockchains must develop a better way of providing authentication to the mining system without the necessity of using large amounts of energy to do it. It is also possible to eventually use processing power from green electricity to conduct the mining.

Blockchains are also under review to be a part of the energy infrastructure. The use will allow energy companies to be decentralized, and a significant reduction in infrastructure and labor is expected. In the event this happens, it is possible that the energy needed to run the system could be attached to the energy producers.

The other option would be to choose proof of stake over proof of work as discussed. There is also the possibility of introducing some other form of investment into the blockchain environment. If you look at the proof of work or the proof of stake, you'll see a dynamic of return-driven activity. To be able to mine under POS criteria, you would have to purchase the coin. To be able to mine under the POW criteria, you would have to invest in the equipment. In either case, you need to invest in something, so that you are, in the end, a stake holder.

The future of blockchains depends on finding the proper balance of the investment required to entice those who wish to advance as miners.

The other problem that is an issue with blockchains such as Bitcoin, is that they are susceptible to companies pooling their hashing power. At first glance, that does not look like a bad thing, until you recall something called the Fifty-one Percent Rule. This rule is all about the number of blocks that are confirmed by the nodes as a percentage of the nodes in the network. It is done by consensus, remember? If 51% of the nodes in a network agree to a block, then that is the block that takes on validity. Anyone having more than 51% of hashing power in a blockchain network should be a red flag. In the case of Bitcoin's blockchain, two companies have 53% of the total hashing power. This is a situation that is ripe for collusion and the cornering of the market in the event such an opportunity should arise.

There are pools that have miners who sign up with

them and they (the administrators) control what happens. This makes me uneasy, and the development of blockchain technology should find a way of taking care of potential problems that arise from this. Besides, pools like this bring tremendous hashing power and promote that power to a point that the difficulty keeps escalating, and so the total power keeps going up just to produce the same output of work. This certainly needs to change.

The next development of the blockchain includes the introduction of AI - Artificial Intelligence, Smart Learning and Machine Learning into the blockchain. Most tech thinkers are excited about the prospect of this combination of two of the next generation's ideas merging.

The development of such a combination is already underway and aside from a few small initiatives, there isn't much to write home about just yet. However, it is pretty certain that within the next five to ten years the AI/Blockchain combination will attain a strong footing in the world, and that will cause the shift in how we live our lives. We are already seeing the Internet of Things changing the way Blockchain is used. We will then see how Big Data is passed and stored, then comes the amalgamation of the technologies that will allow for human beings and the civilization we live in to take the biggest leap yet.

The Internet of Things

If you haven't heard of the IoT, let me just give you a brief review of it. The IoT is about being able connect all equipment and electronics, from your fridge, to your watch, sunglasses, car, toaster—everything. Once you tag all of these things and put them on to the internet, you will be able to control them in any way from anywhere. The Internet of things is designed to connect devices the way social media connects humans (not in the exact same way, but you get the point).

The development of the blockchain technically has the potential to be the platform to connect devices around the world to our own device, without us worrying that it could get hacked. Wearable computing meeting big data will inadvertently land us all in AI territory. Keeping the blockchain (or some more advanced from of it) will move this along faster and further in a shorter period of time.

Blockchain holds the key to being resilient to DDOS attacks and to be able to overcome virus attacks as well as malware issues. As you have seen in other parts of the book, the blockchain can be designed to be impervious to the typical bad actor. It is also easy hide in the cloud, break up the footprint and scatter it across the blockchain.

The internet of things gives as a great way to further the

power of the blockchain to be not only part of the nodes but also the tools which make it impervious to attacks.

Chapter 6

Different Blockchains

There are numerous blockchains across different architectural formats written in numerous programming languages. There are a string of them that you can download from GitHub. There are also blockchains that are varied in nature and objective, and you can get just about anyone who can use some sort of object-oriented programing to create a new blockchain with your objective in mind. Blockchains are the results of programing parameters and processes, not a physical architecture that needs to be built with physical materials.

But for now there are three special blockchains that I want to look at in particular and they all have to do with cryptocurrencies, tokens, and smart contracts. You see, the blockchain beneath the utility layer, as discussed earlier in the book, has some impact on how the blockchain below behaves. If it is just purely currencies like the Bitcoin network, then the mechanisms in place are a little different. If it's for smart contracts then the mechanism is a little different. Finally, if they are meant for tokens, then the structure is just a little different as well.

The currency blockchain can be divided into two main avenues—the bitcoin blockchain and the altcoin blockchain. Most of the blockchains require proof of work for the miners to be able to have a shot at the rewards. Then comes the Ethereum blockchain that is optimized for the transmission of smart contracts. And finally there is the blockchain that manages tokens like that of Filecoin. I will get into the last two in turn, as we have already exhaustively looked at bitcoin earlier.

Blockchain vs Hashgraph

The one thing that should start becoming apparent is that the blockchain is not just attached at the hip to any particular cryptocurrency. Even more than that, the blockchain is not exclusive to Bitcoin. Bitcoin indeed does have its blockchain, but it must be fully understood that the blockchain is totally malleable and can be designed to support almost anything that it needs to on top of it. The blockchain in essence becomes the platform for the application that's added on top of it. In between the application and the platform of the blockchain, are the tokens or cryptos that facilitate the connection and bring monetary or currency value between the two layers.

There have been multiple attempts to 'alter' the role of the blockchain, but in effect what people are really doing is altering the format of the blockchain, not the concept of it. One such situation is the development of the Hashgraph. It is seen as the possible replacement to the current format of the blockchain.

Hashgraph is what they call a distributed ledger with consensus algorithm. They have gone through great pains to not call it blockchain and there is a reason for that. First of all you have to remember that Satoshi never patented the blockchain. The developer of Hashgraph, Leemon Baird, did however, patent Hashgraph. And so if he wanted to patent it, it had to be unique. Therefore, there are a few things he has done to differentiate it which worked really well. But it is in essence another blockchain.

Let me give you an idea of how the Hashgraph system works (since they don't want to call it a blockchain, I shall respect their wishes). Hashgraph does indeed do a few things better than the traditional blockchain. For one, it does not allow the miners to pick and choose which transaction it wants to include in a block. It all has to be included based on the timestamp. If you recall, miners now have the right to choose who they include in their block when it comes to the current bitcoin block. That is obviated in the Hashgraph version. So that's a good thing. In the event you are planning to create your own blockchain, this is probably something that you would want to do.

So the first similarity between Hashgraph and

Blockchain are that they both use a form of gossip protocol. But the difference is that the events are properly timed, and the timed events are not based on what each node says but on the consensus algorithm that runs Hashgraph. So no one can cheat on the time. Also, because there is a fairness protocol in Hashgraph, you get a sense of certainty that a transaction is going to get included in the 'block' regardless of the tip that the transaction participants provide. In bitcoin there is a tipping mechanism where participants provide a fee in the transaction to be given to the miners. Miners have the habit of taking that fee into consideration when they mine the blocks. They tend to leave out transactions that have little or no additional fees and take the ones that have the higher fees.

That's the key difference between Hashgraph and Blockchain. One more thing. Even though they both use gossip protocols, Hashgraph goes one step further and produces gossip about gossip. So that's a little different, in the sense that there are gossips which create events that are not really material events. It means that if A talks to B, that is an event, even if there is no material transaction. The moment A talking to B creates an event, it means that B can now tell C that there was an event of A talking to it. And so now they keep talking about talking. And they do this very rapidly. Smaller packets of events, each containing the time stamp, are passed along. That works great because then what happens is that the Hashgraph is able to keep a timestamp of its own in consensus. When you have a thousand people all with their clocks slightly off, pile over and over again over aen event, there is no way that event is something that happened in the future, so the mistaken (or intentionally altered time stamp) problem is solved in the Hashgraph.

Between the time stamp issue and the way they take transactions it becomes a fairer process. In this case it also becomes a faster process, because the more people there are in the network, the faster the gossip rate becomes and that means that the transaction rate is also faster. Hashgraph claims that it can do more than 50,000 transactions per second compared to Bitcoin's significantly lower number. This is really good news, and if and when you get to develop your own blockchain, these are the things you want to think about.

Not only is Hashgraph fair, it is also cheaper and faster, with the ability to perform more transactions. The developers are also advancing the process to be quantum tolerant.

Blockchain vs Tangle

The Tangle is an interesting version of the blockchain to a certain extent. It takes all the things that work well in the blockchain and improves things by removing some of the aspects that are not needed, while enhancing some of the parts that are needed. For instance, there are no fees and there are no miners. The fact that blockchains use miners makes it impossible to do anything without them, because without miners it is impossible to keep the transactions in blocks and to verify them. However, the idea of being able to tip the miner should not have been something that was ever implemented, as there should not have been fees at all. The block reward should have been more than enough to be able to incentivize the miners. But because it started that way, it is now very difficult to change that and as a result, it is impossible to send microtransactions.

In the case of Tangle there are no blocks, so that means there are no miners. If there are no miners then there are no fees. So far so good. But the most important thing is that it does use distributed consensus. It also offers quantum security which means it can't be ranked when quantum computers become mainstream. Tangle also offers scalability, so that is a very good start. It is something that you might want to consider when you try to develop your own blockchain or wish to fork the code here as well.

Tangle uses what we call a DAG—A Directed Acyclic Graph. It's easy to unpack. Directed means it just goes in one direction and Acyclic tells us that it is linear and noncircular, so it does not go over itself.

The beauty about this sort of blockchain (and again, they do not like being called a blockchain, they call the foundation a tangle) is that they do not need the miners to accumulate transactions; each transaction is verified by two transaction after it and it is done so, randomly.

Let me give you a few terms to play with so that you get a little better understanding of the tangle. The coin that is built on the tangle, by the way, is the IOTA. The tangle doesn't use blocks. Instead it uses transactions that need to be verified by two transactions that happen after it. Each transaction is not called a block, but a site. When a site is created, two sites randomly chosen behind it will verify the first site. Then four other transactions will verify the last two transactions, and that keeps going downstream. The two sites that are verified are chosen by the system using a random algorithm so that no one person can keep approving his or her own transactions. As the sites approve each other, the line of sites form "branches"—so called because they form a crooked or jagged line from the tips to the ends. The tips are the most recent sites that have yet to have any transaction confirm them.

By doing it in this way, what happens is that the scalability rides on itself. The more people there are, the faster the transactions get done. And instead of the blockchain getting jammed because the miners are trying to choose who is paying them more and the fact that each block has a 1MB limit, it is much less chaotic in the world of Tangle than it is in the Bitcoin blockchain.

Chapter 7

Creating a Blockchain

Why would you want to create a blockchain, when everyone is thinking of creating cryptocurrencies? Well the reason you would do that is so that you will have greater freedom in the features that the blockchain can support. Let's look at it this way, blockchains are not just the foundation of cryptocurrencies, they are also the foundation of other distributed applications—DApps.

DApps require platforms and while they can be launched and hosted from servers, they are better used and optimized to work off decentralized nodes. This is where the blockchain comes in.

If you remember, we talked about P2P systems where apps back in the '90s, similar to Napster, would open a port on a computer and allow you to share files in that folder with anyone in that network. Think about that for a minute. A computer in Dubai could connect directly to a computer in Auckland and pull the data needed. In the days of Napster, the files were typically music files, and at that time they were typically pirated music. Let's forget the issue of entertainment piracy, but look at the ability of one computer to store data on another computer. The way it worked back then, the node would only allow access to what was in a particular folder and nothing else. So it was safe.

Now advance that idea a little more. Napster had a registry or a file search that kept track of where all the files were located. so let's say you were looking for a particular song—Metallica's "Enter Sandman." If you typed that into the search, it would give you a list of all the computers that were online at that moment and had that song in its sharing folder. You could pick any one of those, click on it, download the folder and then store it in your folder. You could either leave it there to share with others or you could take it out. But it was that simple.

Now let's take this one step further than P2P concepts. What if I wanted to place a folder that I created in someone else's drive, maybe as a backup? And what if someone wanted to leave something in my drive as well. What would be the problem?

Well the problem would be that I could go and read the content of that file that didn't belong to me. Right? So the way we solve that is to encrypt the data. But before we encrypt it, how about we break that file up into multiple pieces, like a jigsaw puzzle. Imagine taking a picture and breaking it up into tiny jigsaw pieces, encrypting that and then flinging the small pieces across the network of nodes. So once a fragment of that file enters my computer, not only is it encrypted, it is also just a fragment. Even if I were to find some way to decipher the encryption (and this is next to impossible), all I have is just a fragment of jumbled-up pieces of the original file. It will not make any sense to me.

Now imagine that to be a form of monetizing your spare hard drive space, and in return for opening your node to storage, you receive a coin or a token (the amount would vary). The DApp would be the application that lived on top of the blockchain, and the coin would be the one that was between the blockchain layer and the DApp layer. The blockchain would host the immutable record of where all your files are located in a series of hashed public keys, which only your private key could unlock. So in essence your information would be spread across the globe, and in the event of a wipeout somewhere, there would be a backup of the file in a number of places that could just back it up from there.

Now imagine you were planning to write a blockchain that did that. You see, it is a lot more than just one cryptocurrency; it's an entire platform for an ecosystem. So it's a big deal and you can't write it in Java or Python.

For this kind of a project, you can see why you would have to get your own blockchain and code it from scratch. What you can also see is the sudden emergence of an appreciation for a blockchain that only seemed to have been useful to support cryptocurrencies.

Blockchains, while unarguably the backbone of cryptocurrencies, have a lot more potential. They are in essence the core of a decentralized system that allows consensus building and permission. When you pair that with strong cryptography—you could even require SHA512 instead of the current SHA256——then you would see that blockchains become even more robust and longer lasting.

I mention longer lasting because at this point in time, one of the few risks in blockchain integrity comes from the rate of increase in processing power. When they first started mining, it would be done on CPUs in home computers. Then it needed something more powerful because there were more miners. We've seen that the more miners there are, it increases the chances that the solution to the puzzle is found sooner, and so to keep the time at a constant 10 minutes or thereabouts, the program has to increase the difficulty of the solution. This is what spawned harder and harder solution requirements. As it increased in difficulty, miners found more ways to get processing power and even resorted to using GPUs. After that, manufacturers started making ASICS, powerful processors that only perform one operation—hashing. But all this is predicated on the same processor technology. No doubt we have faster processors, but they are still limited by a design ceiling.

But within the next decade, the way we compute is going to change as quantum computing comes online, and what now takes hours to hash, will take seconds to complete. In essence, the ability to mine faster will skew the POW formula as it stands and that could affect the integrity of the blockchain itself, as blocks that do not conform to standards could still be included into the chain and things like double spending could reduce the efficiency of the platform. So for those of you even considering the possibility of building your own blockchain for an application that you envision, do keep this in mind and find a solution for it now so that you don't have to face the problems tomorrow when quantum computing hits the market.

If you are indeed thinking of creating your own blockchain, here are some of the things you need to consider. They are the starting points you need to think about, reflect on, and solve, before you key in your first line of code.

Language

In my experience, the best way to code blockchains is to do it in C++

Here is why. First off, if you are making this open-source, C++ assures you a simple fact: only solid programmers know the best C++ and that makes it a barrier in itself! You really do not want to use a language that a lot of problems could arise from. By using C++ you create a natural barrier.

Secondly, the reason you want to use C++ (and by the way, Bitcoin is based on it), is that it allows powerful resource management—more so than most of the other more recent and hip languages. Remember, C++ is built on the original C programming and still carries a lot of its power in addition to the improvements that were made when C++ was developed. It is also a language that has been around for some time, giving it the necessary street cred to make it work.

Other than that, the power of C++ to harness the necessary resources across the various operating systems that will inevitably be part of the node population, makes it an obvious choice for the language that is the most versatile in overcoming challenges of scalability, control, and responsiveness over a wide-ranging and far reaching landscape.

You should be aware that developing your own blockchain is a lot harder than just forking it from a Git somewhere. In that case, you are just taking someone else's concept and making it your own. There is nothing wrong with that but be aware that it will not contain all the features and attributes that you will want. But nonetheless, it is a good place to start.

Security

The first thing you need to remember is that the best form of blockchain programing and building you can do is the one that is open source if you are planning to have it as widespread and robust as possible. Open source provides you with the widest possible technical ability and it gives you the best path to a robust product. It also gives your competitors a launching platform, but no need to worry about that. The best development projects are done in open source and there is more to just having code that makes something duplicable.

To be open source you have to have mechanisms that deter bad actors and to be self-correcting. You should think of best practices in open source deployment before you can move forward.

Your first goal is security, and this is not done by placing anti-viruses or anti-malware code. No. This is accomplished by the structure of your ideas and implementations concepts that are then conceded. The first thing you need to design before you get coding is how the process of the blockchain will work. Any changes to that post-deployment should be blocked.

You should also consider using something more powerful than SHA256 and keep in mind that quantum computing is around the corner, so your blockchain should be robust enough to keep up with the change in technology. This is not to say that you have to develop something that can withstand any and all technologies yet to be developed. Instead you have to build a system that is modular and has the scalability in terms of loads as well as technology. C++ in open source format with modules is one of the best ways to accomplish this. What's even better is that C++ is one of the easier languages to host a scripting language on top of it. Consider the Arduino microcontroller. It is a small but versatile microcontroller board that is driven by a program called Sketch, but it is actually based on C++. That makes it simple to execute and easy for anyone to learn, yet powerful enough to withstand the harshest environment. Ethereum does this as well. They have a programing language to write the smart contracts on top of the Ethereum blockchain. It's called Solidity and it is really simple to use even if you do not know C or any other language.

Conclusion

That brings us to the conclusion of our endeavor to understand not just the blockchain beneath the bitcoin, but the blockchain in general and the competing ideas that are fast coming up the chain.

The thing that most people fail to understand is that there are certain qualities that make the blockchain a blockchain. It is not a proprietary name and so it is more of a concept and we should not look at it is as being proprietary to Bitcoin. It's like saying that the Internet is only for news or for CNN. The Internet is the platform and everything is built on that. It doesn't matter if HTML is 2.0 or 5.0, it is still the Internet. In the same way, whether it is Hashgraph, Tangle, or Ethereum, it doesn't matter as each is a form of blockchain. They just has different purposes and offer different products. Ethereum looks at smart contracts on top of its blockchain, and that's the feature they chose to focus on. Bitcoin looks at currency, and Tangle looks at the IoT space. They all staked out the application layer and built the blockchain below to do something fantastic that would promote the app on top. But they are all the same. Out of the 1200 over cryptos and tokens in the market today, not all of them possess unique blockchains. In fact some of them have forks of the original Bitcoin blockchain. And that is fine too. But the point is that whether you want to go out and fork your own blockchain or build one from scratch, then you need to clearly define what it is you are doing.

What I can tell you is that this is the wild west of the future of the internet! Just as e-commerce on servers was the "in thing" during the '90s and the '00s, this is the start of the new billionaire's bubble. The question is, which space do you want to get into and what do you want to do? Just merely trading the coins will get you some beer money here and there. But where is it you really want to make an impact?

That is the real question.

Thank You!

Before you go, we would like to thank you for purchasing a copy of our book. Out of the dozens of books you could have picked over ours, you decided to go with this one and for that we are very grateful.

We hope you enjoyed reading it as much as we enjoyed writing it! We hope you found it very informative.

We would like to ask you for a small favor. <u>Could you please take a moment to leave a review for this book on Amazon?</u>

Your feedback will help us continue to write more books and release new content in the future!

Don't Forget to Download Your Bonus:

Bitcoin Profit Secrets

https://dibblypublishing.com/bitcoin-profit-secrets

More Books by

Crypto Tech Academy

- Cryptocurrency Trading: A Complete Beginners Guide to Cryptocurrency Investing with Bitcoin, Litecoin, Ethereum, Altcoin, Ripple, Dogecoin, Dash, and Others
- Cryptocurrency Mining: A Complete Beginners Guide to Mining Cryptocurrencies, Including Bitcoin, Litecoin, Ethereum, Altcoin, Monero, and Others

Made in the USA
San Bernardino, CA
22 March 2018